The M26 Pershing and Variants

T26E3/M26 • M26A1 • M45 • M46/M46A1

Troy D. Thiel

Schiffer Military History
Atglen, PA

Photo Credits
Admiral Nimitz Museum.
United States Army
United States Marine Corps
Mr. Clarence Smoyer
Patton Museum
Wright Museum of American Enterprise

Cover artwork by David Pentland,
"Retreat from the Yalu",
courtesy of the artist.

Book design by Robert Biondi.

Copyright © 2002 by Troy D. Thiel.
Library of Congress Catalog Number: 2001096784.

All rights reserved. No part of this work may be reproduced or used in any forms or by any means – graphic, electronic or mechanical, including photocopying or information storage and retrieval systems – without written permission from the copyright holder.
"Schiffer," "Schiffer Publishing Ltd. & Design," and the "Design of pen and ink well" are registered trademarks of Schiffer Publishing, Ltd.

Printed in China.
ISBN: 0-7643-1544-7

We are always looking for people to write books on new and related subjects. If you have an idea for a book, please contact us at the address below.

Published by Schiffer Publishing Ltd.
4880 Lower Valley Road
Atglen, PA 19310
Phone: (610) 593-1777
FAX: (610) 593-2002
E-mail: Schifferbk@aol.com.
Visit our web site at: www.schifferbooks.com
Please write for a free catalog.
This book may be purchased from the publisher.
Please include $3.95 postage.
Try your bookstore first.

In Europe, Schiffer books are distributed by:
Bushwood Books
6 Marksbury Ave.
Kew Gardens
Surrey TW9 4JF
England
Phone: 44 (0)208 392-8585
FAX: 44 (0)208 392-9876
E-mail: Bushwd@aol.com.
Free postage in the UK. Europe: air mail at cost.
Try your bookstore first.

The M26 Pershing and Variants

The M26 Pershing was the product of a prolonged development that began in 1942. Production and deployment of the Pershing was delayed by conflicting opinions on the requirements for a tank to replace the M4 Sherman. Reaching the ETO in the spring of 1945, the Pershing proved to be a sound, if not perfect, design. The M45, a close support variant armed with a 105mm howitzer, was produced in very limited numbers but saw no action during World War II.

Postwar cuts in defense spending prevented the adoption of any new tanks, and in the late 1940s 800 Pershings were rebuilt with improved drivetrains and guns, and adopted as the Medium Tank M46 Patton. In addition, a number of otherwise standard M26s were rearmed with the new 90mm of the M46 and redesignated M26A1. With the North Korean invasion of the South in 1950, the U.S. went to war with the M26 and M26A1 Pershings, M45, M46 Patton, and a roughly equal number of Shermans. The Pershing and its variants performed well in Korea and played a vital role in stemming the advance of the NKPA early in the war. A final variant, and the last rebuild of the World War II era Pershing, was the M46A1, which incorporated many features of the new M47 Patton medium tank. Like the M4 Sherman, the M26 Pershing and its variants remained in service with the U.S. military until the late 1950s. In addition to service with the U.S. military, the M26, M26A1, and M46 also saw service with several NATO armies.

Although it never replaced the Sherman, the Pershing proved to be just as long-lived and versatile as its smaller stablemate. Arriving on the scene too late to have an appreciable effect on the outcome of World War II, the Pershing and its variants were the best tanks in the U.S. arsenal during the first years of the Cold War. The Pershing is also the direct ancestor of the later M48 and M60 tanks.

The T26E3/M26 Pershing

After a long and troubled development, the T26E3 heavy tank finally entered production at the Fisher Tank arsenal in November 1944, and at the Chrysler Detroit Tank Arsenal in March, 1945. By the end of the war T26E3 had been adopted as the M26 Pershing. When production ended in late 1945 a total of more than 2,000 tanks had been built by Fisher and Chrysler.

The Pershing was armed with the M3 90mm gun, which was able to penetrate up to 6.1 inches of armor at 2,000 yards using HVAP ammunition, with a muzzle velocity of 3,350 feet per second. This was superior to the performance of the 88mm KwK36 of the Tiger I but slightly inferior to the 75mm KwK42 of the Panther. During the course of its production several changes were made in the design of the M26. The early T26E3 was equipped with a 400 cfm rotoclone blower to provide ventilation. The blower was mounted behind a rounded bulge in the glacis plate, between the driver's and bow gunner's hatches, and featured a single intake port on either side of the housing. The 400 cfm blower was later found to provide inadequate ventilation and its rounded housing proved to be a weak point in the glacis plate. Beginning with tank number 550 at the Fisher Arsenal and number 235 at the Chrysler arsenal, the 400 cfm rotoclone blower was replaced by a 1,000 cfm version. The early rounded blower housing was replaced with a thicker, flat surfaced housing with two intake ports on either side, which provided much better ballistic protection. Another early design feature was a pair of rotating periscopes mounted on the hull roof, on either side of the blower housing. These periscopes were found to be vulnerable to enemy rounds that might be deflected downward after a hit on the mantlet, and were deleted from the design at the same time as the 400 cfm blower. Many early tanks had these periscopes removed and their mounting holes welded over. The design of the travel lock for the 90mm gun was also modified during production. Early tanks featured a travel lock mounted directly to the exhaust. This was replaced by a stronger design mounted on the rear of the hull. The travel locks of most early production tanks were modified to the later style. The T26E3 was originally equipped with the T81 24 inch single pin track, but an improved 23 inch double pin track based on that of the HVSS-equipped Shermans was later adopted. The T80E1 track was produced with both all steel and rubber chevron links. Both the T81 and T80E1 tracks remained in use into the 1950s.

Although it suffered from a lack of range compared to the Sherman, and was underpowered for its weight, the T26E3/M26 proved to be popular with crews when introduced in early 1945. In combat the Pershing proved to be a match for both the Panther and Tiger I. In turn, Pershings were knocked out by the 88mm guns of the Tiger I and Nashorn. In Korea the Pershing performed very well against the T34/85s of the North Korean People's Army, although they were vulnerable to the Soviet supplied 57mm, 76mm, and 85mm guns.

The M45

Envisioned as a close support tank to compliment the M26, the T26E2 was armed with the same 105mm howitzer as the M4E5 Sherman. With the exception of the main gun, the T26E2 was basically identical to the late production Pershing. The surrender of Germany led to the cancellation of the Fisher Tank Arsenal's contract before production started, and the T26E2 saw very limited production at Chrysler's Detroit Tank Arsenal.

A total of 185 tanks were produced in the last six months of 1945. After World War II the T26E2 was standardized as the Medium Tank M45 and saw limited service during the Korean War, where the M4E5 Sherman was more commonly used in the support role.

The M26A1

In the late 1940s a number of Pershings were re-armed with the M3A1 90mm gun. The new gun can be identified by its bore evacuator and a single baffle muzzlebrake. The travel lock for the 90mm was relocated from the exhaust to the rear deck, but the M26A1 was otherwise mechanically identical to the M26. Like the M26, the M26A1 saw service with the U.S. military during the Korean War, and was also used by NATO countries, including Belgium and Italy.

The M46 Patton

In the postwar years a tight budget prevented the adoption of a new tank, so an attempt was made to update the Pershing and correct many of the original design flaws. The result was the Medium Tank M46 Patton. The underpowered 500hp Ford GAF powerplant was replaced by an 810hp Continental AV-1790-3 aircooled engine and General Motors crossdrive transmission. Other improvements included the addition of disc brakes and the replacement of the M26's steering levers with a joystick. The M3 90mm gun was replaced with an improved version, the M3A1, which featured a bore evacuator and single-baffle muzzlebrake. The rear hull was redesigned with the exhaust being routed through two large mufflers on the rear fenders, and the travel lock for the 90mm was relocated to the rear deck. Also, a track-tensioning wheel was added between the last roadwheel and drive sprocket. The M46 was a great improvement over the original M26 design. The M46 proved to be well suited to the mountainous terrain of Korea, due to its improved horsepower to weight ratio and handling characteristics. During the Korean War a number of M46s were further modernized, resulting in the M46(New), which was externally identical to the M46 but was equipped with the drivetrain and control systems of the M47 Patton, which was just entering service. The M46(New) was later redesignated as the M46A1 to avoid confusion between the models, and both Pattons saw service in Korea alongside the M26, M26A1, and M45. Like the M26 and M26A1, the M46 also served with NATO countries until replaced when stocks of M47 Pattons became available.

Technical Data

T26E3/M26, M26A1

Drivetrain and Suspension

Motor	Ford GAF V8, gasoline
Bore & Stroke	5.4X6
Displacement	1,100 cubic inches
Maximum Power	500 hp at 2600 rpm
Cooling	liquid
Batteries	2 12 volt DC in series
Generator	1 150 amp, belt driven
Transmission	Torqmatic 3 speeds forward, 1 reverse
Suspension	Torsion bar
Running Gear	6 dual roadwheels and 5 dual return rollers per track, front idler, rear drive
Track width	24 inches (T81), 23 inches (T80E1)

Armor, hull

Front, upper	4.0 inches
Front, lower	3.0 inches
Side, front	3.0 inches
Side, rear	2.0 inches
Rear, upper	2.0 inches
Rear, lower	0.75 inches
Top	0.875
Floor, front	1.0 inches
Floor, rear	0.5 inches

Armor, turret

Mantlet	4.5 inches
Front	4.0 inches
Sides	3.0 inches
Rear	3.0 inches
Top	1.0 inches

General Data

Height	109.4 inches
Length	340.5 inches
Width	138.3 inches
Weight, empty	84,850 lbs
Weight, combat	92,355 lbs
Fording depth	48 inches
Ground clearance	17.2 inches
Top Speed	30 miles per hour
Range	100 miles
Fuel capacity	183 gallons
Crew	5
Armament	1x 90mm M3 (70 rds), 1x .50 MG (550 rds), 2x .30 MG (5,000 rds)

M45

Drivetrain and Suspension

Motor	Ford GAF V8, gasoline
Bore & Stroke	5.4x6
Displacement	1,100 cubic inches
Maximum Power	500 hp at 2600 rpm
Cooling	liquid
Batteries	2x12 volt DC in series
Generator	1 150 amp, belt driven
Transmission	Torqmatic 3 speeds forward, 1 reverse
Suspension	Torsion bar

Running Gear .... 6 dual roadwheels and 5 dual return rollers per track, dual front idler, rear drive sprocket
Track width .... 24 inches (T81), 23 inches (T80E1)

Armor, hull

Front, upper .... 4.0 inches
Front, lower .... 3.0 inches
Side, front .... 3.0 inches
Side, rear .... 2.0 inches
Rear, upper .... 2.0 inches
Rear, lower .... 0.75 inches
Top 0.875
Floor, front .... 1.0 inches
Floor, rear .... 0.5 inches

Armor, turret

Mantlet .... 8.0 inches
Front .... 5.0 inches
Sides .... 5.0 to 3.0 inches
Rear .... 2.5 inches
Top .... 1.0 inches

General data

Height .... 110.9 inches
Length .... 256.7 inches
Width .... 138.3 inches
Weight, empty .... 86,000 lbs
Weight, combat .... 93,000 lbs
Fording depth .... 48 inches
Ground clearance .... 17.2 inches
Top Speed .... 30 miles per hour
Range .... 100 miles
Fuel capacity .... 183 gallons
Crew 5
Armament .... 1x105mm M4 (74 rds), 1x .50 MG (550 rds), 2x .30 MG (5,000 rds)

M46, M46A1

Drivetrain and Suspension

Motor .... Continental AV-1790-5A (M46), AV-1790-5B (M46A1) V12, gasoline
Bore & Stroke .... 5.75 x 5.75
Displacement .... 1,791.7 cubic inches
Maximum Power .... 810 hp at 2,800 rpm
Cooling .... air
Batteries .... 4x12 volt DC, 2 sets in series connected in parallel
Generator .... 1 150 amp
Transmission .... cross-drive cd-850-3 (M46), cd-850-4 (M46, M46A1) 2 speed Forward, 1 reverse
Suspension .... Torsion bar
Running Gear .... 6 dual roadwheels, 5 dual return rollers, and one dual tensioning wheel per track. Front mounted dual idler, 1 dual tension idler between last roadwheel and rear mounted drive sprocket
Track width .... 23 inches (T80E1, T80E4, T84E1)

Armor, hull

Front, upper .... 4.0 inches
Front, lower .... 3.0 inches
Side, front .... 3.0 inches
Side, rear .... 2.0 inches
Rear, upper .... 2.0 inches
Rear, lower .... 0 .75 inches
Top .... 0.88 inches
Floor, front .... 1.0 inches
Floor, rear .... 0.5 inches

Armor, turret

Mantlet .... 4.5 inches
Front .... 4.0 inches
Sides .... 3.0 inches
Rear .... 3.0 inches
Top .... 1.0 inches

Technical Data

General Data

Height	125.1 inches
Length	333.6 inches
Width	138.3 inches
Weight, empty	92,000 lbs
Weight, combat	97,000 lbs
Fording depth	48 inches
Ground clearance	18.8 inches
Top Speed	30 miles per hour
Range	80 miles
Fuel capacity	232 gallons
Crew	5
Armament	1x 90mm M3A1 (70 rds), 1x .50 MG (550 rds), 2x .30 MG (5,500 rds)

Sources

Hunnicutt, R. P. *Patton: A History of the American Main Battle Tank*. Novato, California: Presidio Press, 1984.

Hunnicutt, R. P. *Pershing: A History of the Medium Tank T20 Series*. Bellingham, Washington: Feist Publications, Inc., 1996.

Chapter 1:
M26 Pershing

Fisher Tank Arsenal T26E3, serial number 221, on display at the Patton Museum at Fort Knox, Kentucky. This tank was refurbished at the Detroit arsenal in 1948 and saw service with the Italian Army. It was returned to the U.S. in 1977 for use as a gunnery target during A-10 "Warthog" aircraft tests, and later recovered from Eglin AFB, Florida by the 2BN/70AR. The markings represent the 70th Tank Battalion during the Korean War. The tiger face marking is one of several variations seen on U.S. tanks during the Korean War. (Author's Collection)

Opposite: Three quarter view of the T26E3 at the Patton Museum. (Author's Collection)

Rear view of the Patton Museum T26E3. The gun travel lock has been removed from the exhaust housing and replaced by one welded directly to the hull. The original travel lock mounts can be seen projecting from either side of the exhaust. The bracket below the M26 marking held the towing pintle when not in use. The large flat bracket on the right side of the hull is a mount for a phone that allowed infantrymen to communicate with the tank's crew. The brackets on the turret rear held the .50 cal. Browning when dismounted. (Author's Collection)

Opposite: Detail of the T26E3/M26 suspension and later T80E1 double pin track with rubber chevron blocks. (Author's Collection)

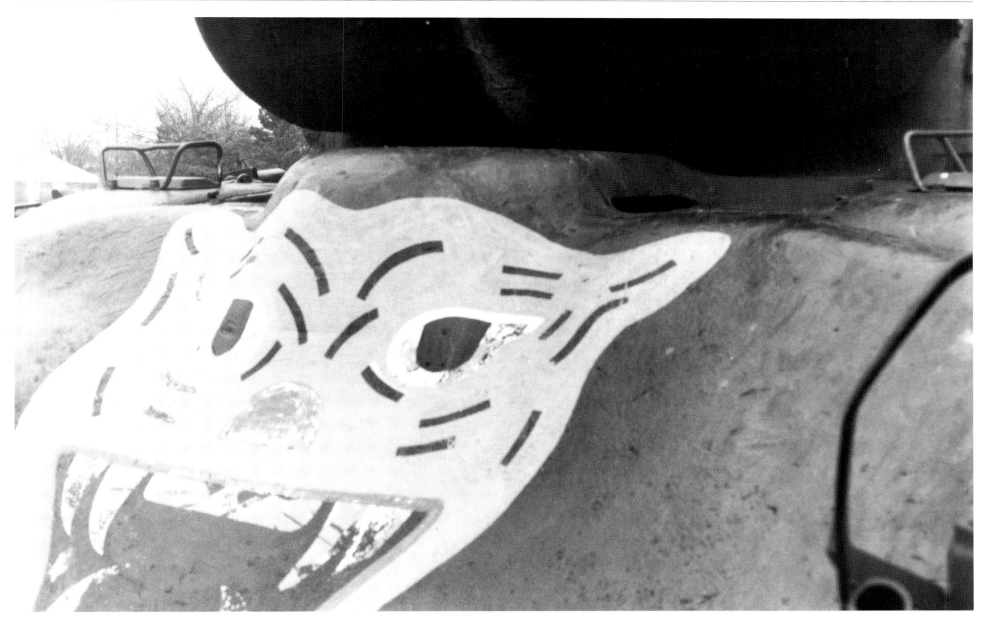

Opposite: Detail of the T26E3/M26 rear-mounted drive sprocket. (Author's Collection)

Detail of the 440 cfm Rotoclone blower housing of an early production T26E3/M26. The 400 cfm blower can be identified by the rounded housing and single intake on each side. The hull periscopes on either side of the housing have been replaced with welded plugs as they were found to be vulnerable to rounds ricocheting downward from the mantlet. (Author's Collection)

Detail of the later production 1,000 cfm Rotoclone blower. Note the thicker, flat housing and dual intakes on each side. The hull periscopes on either side of the housing were deleted from the design when the new blower was adopted. (Author's Collection)

Opposite: A T26E3 throws up a shower of mud during a test run at Aberdeen Proving Ground, 1945. (U.S. Army)

The M26 Pershing and Variants

An excellent rear view of a T26E3. Note the early tall gun travel lock mounted on the exhaust. (U.S. Army)

Opposite: Pershings of the 14th Tank Battalion, 9th Armored Division, wait for orders in a field on the outskirts of Vettweiss, Germany, 1 March, 1945. Note the single pin T81 tracks and full sandshields. (U.S. Army)

Opposite: A T26E3 of CO "C", 19th Tank Battalion, 9th Armored Division near Thum, Germany, passing an M32B1 TRV fitted with mine rollers, March 1945. (U.S. Army)

T26E3 serial number 26 of CO "E", 32nd Armored Regiment, 3rd Armored Division, on the outskirts of Cologne, Germany, 5 March, 1945. This series of photographs, by U.S. Army photographer Jim Bates, and film footage document the destruction of a Panther in front of the Cologne Cathedral by this T26E3 on 6 March, 1945, and were provided to the author by the T26E3 gunner, Mr. Clarence Smoyer. This T26E3 and crew later knocked out a second Panther near Paderborn. (U.S. Army)

The T26E3 firing at German snipers concealed in damaged buildings. (U.S. Army)

Opposite: Advancing through the rubble-filled streets of Cologne. A Panther is loose in the center of Cologne and the T26E3 has been ordered to hunt it down. (U.S. Army)

Advancing cautiously, the T26E3 stalks the Panther in the streets near the cathedral, which is visible in the distance. (U.S. Army)

Opposite: Entering an intersection near the cathedral, the T26E3 comes face to face with the Panther at point blank range. The T26E3 fires on the move, the 90mm's muzzleblast kicking up a cloud of dust. (U.S. Army)

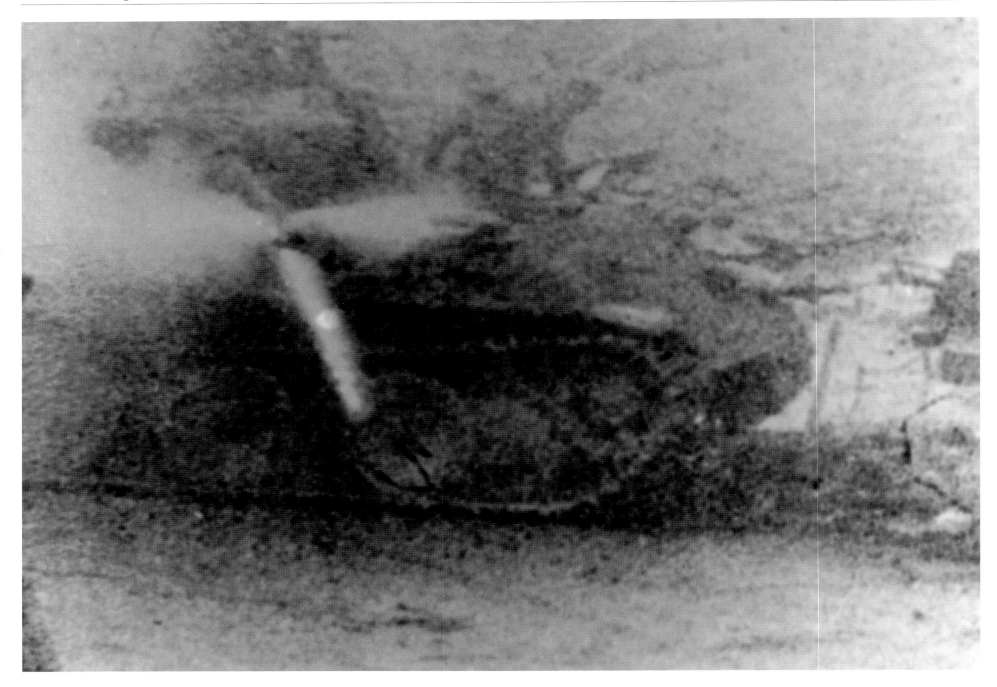

Opposite: A direct hit!. The hole made by the 90mm round can be seen as a glowing circle on the hull just above the Panther's tracks. The driver bails out as the 75mm gun fires. (U.S. Army)

The T26E3 halts and fires two more rounds and the Panther goes up in flames. Two of the 90mm rounds passed through both sides of the Panther's hull, proof of the effectiveness of the Pershing's M3 gun. (U.S. Army)

Left: Examining the kill. U.S. troops inspect the burned out Panther. The soldier armed with the Mauser rifle is the T26E3 assistant gunner, Pfc. John Dereggi. His leather helmet is French-issue for armor troops and was sometimes worn by US tankers. (U.S. Army)

Opposite: An M26 Pershing of the 63rd Heavy Tank Battalion crosses the Rhine at Bruhl, Germany, during a postwar training exercise. Note the unusual placement of the unit insignia on the glacis plate and the star on the lower plate. (U.S. Army)

Opposite: A crewman displays the 90mm round of the M3 gun during a postwar training exercise. This M26 is equipped with the T80E1 track and the travel lock has been relocated to the rear hull. Note the towing cable stowed on the hull rear. (U.S. Army)

An M26 flanked by an M24 and several M8 armored cars near Taegu, Korea, July 1950. This was one of the first three Pershings, under the command of Lt. Samuel R. Fowler, that were rushed to Korea to counter the North Korean T-34s. Found in an ordnance depot in Japan and hurriedly rebuilt, all three Pershings were later abandoned after overheating due to having been fitted with incorrect fan belts. (U.S. Army)

Lt. Fowler's tank crews test firing their guns. Note the WW2 leather and fiber helmets. (U.S. Army)

Opposite: The 90mm of one of Lt. Fowler's Pershings raises a cloud of dust during test firing. (U.S. Army)

The M26 Pershing and Variants

The T26E3/M26 Pershing

Opposite: A column of Marine Corps Pershings in Korea. (USMC)

Pershings firing on enemy positions across the Naktong river, August. (U.S. Army)

An officer directs the fire of a Marine Corps M26 near Yongsan, August 1950. Note the 90mm shell casings on the rear deck that have been tossed out of the open pistol port. (U.S. Army)

Opposite: Infantrymen pass an M26 that has been disabled by a mine in a Korean village. Note empty 90mm casing in the road. (U.S. Army)

Opposite: A Marine Corps Pershing and M4E5 Sherman open fire across a rice paddy near Yongsan, Korea. Note the bulldozer blade mounted on the Sherman. M26 units in Korea often included 105mm howitzer-armed Shermans for fire support. (U.S. Army)

An M26 fires on North Korean troops clearing a minefield, August 1950. The vehicles in the foreground are knocked-out T-34/85s. The vehicle in the road in front of the Pershing is a knocked-out SU-76. (U.S. Army)

Opposite: Directed by a military policeman, a Pershing of the 73rd Tank Battalion moves onto a Korean road, August 1950. Note the equipment stowed on the turret sides and the name MARGARET painted on the blower housing. The bow machinegun is fitted with a protective canvas cover. This tank is equipped with T80E1 tracks. (U.S. Army)

Another Pershing of the 73rd Tank Battalion, named ALICE on the blower housing and fender storage box. This tank has retained its full set of sandshields. (U.S. Army)

Marine Corps Pershings move past a group of North Korean prisoners in September 1950. (U.S. Army)

Opposite: M26s and M4A3s of the 1st marine Division being resupplied in Korea, September 1950. The Pershing in the foreground is equipped with the T81 single pin track. (USMC)

Opposite: An M26 of the 1st Marine Division on the heights above the Naktong river, September 1950. Note the machinegun ammunition cans stored on the fender, and the water cans and 90mm casings on the rear deck. (USMC)

Marines supported by Pershings and an M4E5 on patrol in Korea. (USMC)

Marines climb aboard Pershings in preparation for an advance during the Korean War. Note the M4E5 Sherman support tank on the left. (U.S. Army)

Opposite: Marine Corps Pershings fight their way past enemy strong points near Seoul. Note the canteens and cups hanging from the rear of the turret and USMC camouflage shelter half rolls on the rear deck. This M26 has had the travel lock relocated to the hull rear several inches off center. (USMC)

Opposite: Pershings of "A" and "D" CO, 1st Tank Battalion, deploy to attack an enemy-held village. (USMC)

An M26 carrying troops of the 9th Infantry Regiment advances to halt an enemy attempt to cross the Naktong River, 3 September, 1950. (U.S. Army)

Pershings of the 1st Marine Division on the edge of Kimpo airfield, September 1950. The aircraft in the background is a North Korean Yak. (USMC)

Opposite: Marine Corps Pershings in a hull-down position await an enemy attack, 2 September, 1950. (USMC)

Opposite: The crew of an M26 examines a map somewhere in Korea. This tank is equipped with the T81 track and the flat-faced housing of the 1,000 cfm Rotoclone blower is clearly visible in this photo. (U.S. Army)

US Marines take cover behind an M26 near Hongchon, May 1951. Note the Russian-made M-44 carbine carried by the Marine on the left. (USMC)

The M26 Pershing and Variants

Parka-clad Marines supported by Pershings advance on the Central Korean Front, winter 1951. The fact that the tanks are traveling on the ridgeline indicates they do not expect enemy action. (USMC)

Chapter 2:
M45

M45s and crews during a postwar training exercise. (Admiral Nimitz Museum)

An M45 of the 6th Tank Battalion, 24th Infantry Division crossing the Kumho River during the advance against North Korean positions along the Naktong River, September 1950. Note the full sandshields on the lead tank. (U.S. Army)

Opposite: Another M45 of the 6th Tank Battalion crossing the Kumho River, providing a good view of the M45s heavy mantlet and gun. Note also the stowage on the turret rear and the mix of tanker and infantry steel helmets among the crew. (U.S. Army)

Chapter 3: M26A1

An M26A1 on display at Ft. Polk, Louisiana. It is an early production Pershing with the 400 cfm rotoclone blower and has all of the modifications common to the M26A1. Although not visible here, the periscopes on either side of the blower have been removed and their holes filled in. It has been rearmed with the M3A1 90mm with bore evacuator. The muzzle brake, which has been screwed on at an angle, has been machined down to a single baffle. The tracks are the T80E1 rubber chevron type. (Author's Collection)

Opposite: Rear view of the Ft. Polk M26A1 showing the travel lock for the 90mm in its new location on the rear deck. The former mounting points for the travel lock are visible on either side of the exhaust, below the upper bolts. The tow pintle is stored in its bracket to the left of the exhaust. (Author's Collection)

M26A1

Opposite: Detail of the stowage rack on the right side of the turret of the Ft. Polk M26A1. The presence of this rack is a key to distinguishing the M26A1 from the M46 when the rear of the hull cannot be seen, as it was usually removed from Pershings that were rebuilt as M46s. (Author's Collection)

Rear view of an M26A1 on display at Ft. Knox, Kentucky. An early Chrysler produced Pershing, serial number 1947, it was rearmed with the M3A1 90mm at Rock Island Arsenal in 1952. The tracks are the T80E1 with badly worn rubber chevrons. The box on the right rear of the hull contained a telephone, which was connected to the tank's interphone system, allowing infantry to communicate to the crew when the tank was "buttoned up." Formerly in service with the French Army, it was returned to the US in 1977 for use as a gunnery target. (Author's Collection)

A pair of M26A1 Pershings advance through the streets of Seoul, South Korea. 29 September, 1950. (U.S. Army)

Opposite: M26A1 Pershings of the 1st Marine Division leading a convoy during the Korean war. Both tanks are equipped with the T80E1 steel track. (USMC)

Opposite: An M26A1 of the 1st Marine Division opens fire on an enemy position. Note the road wheel in the stowage rack on the side of the turret. 27 November, 1951. (USMC)

An M26A1 of the 70th Tank Battalion passes an M4A3 Sherman as it tows a jeep across a river north of Chunchon, Korea during April 1951. (U.S. Army)

Chapter 4: M46 Patton

US Marine Corps M46 medium tanks advance during a Korean winter. (USMC)

Opposite: M46 Pattons of A Co., 1st Marine Tank Battalion firing from prepared positions in Korea. The tanks have been backed into pits to increase the elevation of their 90mm guns for use as indirect fire artillery. Note the stack of 90mm rounds ready for use and their cardboard shipping containers. This photo gives a good view of the redesigned rear deck and fender mounted mufflers of the M46. (USMC)

Opposite: An M46 of the 3rd Infantry division tows a crippled mate during the UN withdrawal from Uijongbu, Korea. 27 April, 1951. (U.S. Army)

Infantrymen of the 24th Division and M46 Pattons of the 6th Tank Battalion, 1st Cavalry Division, prepare to attack Chinese forces as Task Force Byorum in the Kumhwa-Sangyang-Ni sector. 21 September, 1951. (U.S. Army)

An M46 of the 6th Tank Battalion, 1st Cavalry Division, advances on Chinese positions in the Kumhwa-Sangyang-Ni sector. 21 September, 1951. (U.S. Army)

Opposite: An M46 of Task Force Byorum opens fire on Chinese troops. Note the cloud of dust raised by the muzzleblast of the 90mm gun. 21 September, 1951. (U.S. Army)

Opposite: M46s of the 64th Tank Battalion ford a river during a probe into communist-held territory. (U.S. Army)

M46 Pattons of the 65th Infantry regiment, 3rd Infantry Division, advance during heavy fighting on the central front. 17 July, 1953. (U.S. Army)

From its position on Hill 355, an M46 of the Heavy Tank Company, 7th Infantry Regiment, 3rd Infantry Division, fires at Chinese positions on Hill 217. This is a good illustration of the type of fighting faced by US tank crews in the latter stages of the Korean War. January, 1952. (U.S. Army)

Opposite: Dug in atop a hill overlooking the UN delegates base camp, an M46 of Co. C, 73rd Tank Battalion, stands guard over the Panmunjom Road. 15 March, 1952. Note the three kill rings painted on the main gun. (U.S. Army)

M46 Patton

The M26 Pershing and Variants

Opposite: A badly damaged M46, knocked out by an anti-tank gun while supporting the 3rd Infantry Division, is loaded aboard an M25 tank transporter. 6 June, 1951. (U.S. Army)

Loaded aboard the transporter, the M46 is prepared for removal to rear area ordinance shops. 6 June, 1951. (U.S. Army)

Opposite: An excellent study of a US Marine Corps M46 during the Korean War. It is equipped with the T80E1 double pin track. Note the canvas covers over the machineguns and mantlet. The crewmen are wearing the M1941 tanker jacket. (USMC)

M46 Pattons of Co. A, 6th Tank Battalion, 24th Infantry Division take up firing positions in the snow. Note the elaborate tiger face paint schemes, complete with claws painted on the front fenders, and the mounting of both .30 and .50 caliber machineguns on the turrets. 6 March, 1951. (U.S. Army)

Accompanied by an M32 recovery vehicle, an M46 Patton joins infantrymen of the 24th Infantry Division in the withdrawal to better defensive positions near Yong Pong Chon, Korea. 23 April, 1951. (U.S. Army)

Opposite: Raising a cloud of snow, an M46 fires on enemy positions in support of the 19th RCT near Song Sil-li, Korea. 10 January, 1952. (U.S. Army)

Opposite: An excellent frontal view of an M46 modified for use by the 1st Loudspeaker and Leaflet Company, Psychological Warfare Section, 8th US Army. The tank is equipped with the T80E1 double pin tracks with all-steel links. Note the single baffle muzzlebrake that has been machined down from the original double baffle M26 design. 5 June, 1952. (U.S. Army)

A rear view of the same tank clearly showing the details of the redesigned rear hull of the M46, including track tensioning wheel between the drive sprocket and last road wheel. 5 June, 1952. (U.S. Army)

A view of the left side of the psy-war M46. 5 June, 1952. (U.S. Army)

An M46 carrying infantrymen on a high speed test run in Korea. (U.S. Army)

Service troops of the 65th Infantry Regiment refuel an M46A1 Patton in Korea. The M46A1 was externally identical to the M46 but was equipped with the drivetrain and control systems of the M47 Patton. 19 June, 1952. (U.S. Army)

NOTES

NOTES

Also from the Publisher

Germany's First Ally: Armed Forces of the Slovak State 1939-1945. *Charles K. Kliment & Bretislav Nakládal.* The Slovak State was born under the auspices of Hitler's Third Reich and became its first ally on September 1, 1939, when it took part in the invasion of Poland. The Slovak army inherited its weapons, equipment, training manuals and its doctrine from the defunct Czechoslovak Army. Though hampered by a shortage of specialists in its air force, armored units and artillery, it managed to field several division-sized units and sustain them during the initial three years of combat on the Eastern front. This book describes the composition, dislocation and equipment of all branches of the Slovak army and its operational history through the war years.
Size: 8 1/2" x 11" • 250+ photos, maps, color plates • 208 pp.
ISBN: 0-7643-0589-1 • hardcover • $39.95

Rommel in the Desert. *Volkmar Kühn.* The complete story of Rommel in North Africa is told in this large, detailed volume. The reader is taken through the initial formation of the Afrika Korps, the stunning early victories, the battles throughout Cyrenaica, and the race across Libya. Then the defeat at El Alamein and the long retreats to Tripoli and Tunis are explored in full. Every tactical aspect of the North African campaign is examined, with detailed maps of the areas discussed. Over 150 photographs show the men that fought the equipment used, and the conditions that prevailed in the Western Desert.
Size: 8 1/2" x 11" • 224 pp.
ISBN: 0-88740-292-5 • hard cover • $35.00

Rommel and the Secret War in North Africa: Secret Intelligence in the North African Campaign 1941-43. *Janusz Piekalkiewicz.* Janusz Piekalkiewicz chronicles the British Secret Inteligence Service (SIS) and the "Ultra" secret, and its effect on the campaign in North Africa during the Second World War. On the German side, Rommel also knew the use of intercepted enemy messages - an awareness that the British were able to gain only much too late and after many sacrifices.
Size: 7" x 10" • over 220 phtographs, maps • 240 pp.
ISBN: 0-88740-340-9 • hard cover • $29.95

Panzertruppen: The Complete Guide to the Creation & Combat Employment of Germany's Tank Force • 1933-1942. *Thomas L. Jentz.* This book provides detailed answers to questions related to how German tankers fought in World War II. The content of this book is derived solely from original records consisting of war diaries, reports, and technical and tactical manuals written during the war. The story is told as recorded by those responsible for decisions in developing the Panzertruppen and by those who fought in the Panzers. This first volume presents the offensive phase up to October 1942. A second volume is planned that will cover the defensive phase to the end of the war.
Size: 8 1/2" x 11" • over 200 b/w photos • 288 pp.
ISBN: 0-88740-915-6 • hard cover • $49.95

Panzertruppen: The Complete Guide to the Creation & Combat Employment of Germany's Tank Force • 1943-1945. *Thomas L. Jentz.* This companion volume presents how the Panzertruppen fought during their defensive struggle with details on the units, organizations, types of Panzers, and tactics.
Size: 8 1/2" x 11" • 47 b/w photographs, maps, charts • 304 pp.
ISBN: 0-7643-0080-6 • hard cover • $49.95

Red Army Tank Commanders: The Armored Guards. *Colonel Richard N. Armstrong.* Tank and mechanized forces spearhead Red Army operations from the gates of Stalingrad to the center of Berlin. This new book profiles Six Soviet commanders who rose to lead six tank armies created by the Red Army on the eastern front during the Second World War: Mikhail Efimov Katukov, Semen Ill'ich Bogdanov, Pavel Semenovich Rybalko, Dmitri Danilovich Lelyushenko, Pavel Alekseevich Rotmistrov, and Andrei Grigorevich Kravchenko. Each tank commanders' combat career is examined, as is the rise of Red Army forces, and reveals these lesser known leaders and their operations to western military history readers.
Size: 6" x 9" • 15 b/w photos, maps • 476 pp.
ISBN: 0-88740-581-9 • hard cover • $24.95

On the Rear Cover

TOP LEFT: An early production T26E3/M26 Pershing, serial number 27, on display at the Wright Museum of American Enterprise in Wolfeboro, New Hampshire. This Pershing served with the 14th Tank battalion, 9th Armored Division and participated in the capture of the Ludendorf Bridge at Remagen, Germany in 1945. (Courtesy of the Wright Museum)

BOTTOM LEFT: Another view of the Wright Museum Pershing. The tracks are the T80E1 double pin type with rubber chevrons. The T80E1 track was based on those of the HVSS-equipped Sherman tank. The boxes on the fender are for the storage of tools and the personal belongings of the crew. The large rack on the turret is for spare roadwheels but was often used to carry water cans and other equipment. This rack was removed from Pershings rebuilt as M46 Pattons. (Courtesy of the Wright Museum)

TOP RIGHT: A rear view of the Wright Museum Pershing. The markings on the fenders indicate that this tank is from Company "A" of the 14th Tank Battalion, 9th Armored Division. The exhaust-mounted travel lock for the 90mm is visible on the rear hull, below the stowed towing cable. (Courtesy of the Wright Museum)

BOTTOM RIGHT: The Wright Museum Pershing, showing the spare track links stowed on the turret side. The length of the M3 90mm gun is evident from this photo. (Courtesy of the Wright Museum)